Book 20
Let There Be Light

Contents

Light - keeper of the night.

Eric Einstein - he is keeper of the light.

Let's Stop For A Chat

Do you know...
- where most light comes from?
- how light is used?
- what sources of light there are?

Chapter 1 What's so great about light?

We need light to see things. If there is no light, we cannot see anything.

Most of the light on earth comes from the sun. The sun is a **source** of light. A source of light is anything that gives off light energy.

Let's see what happens when we switch off the sun?

Eric's experiment.

4

lamplight

city lights

firelight

moonlight

candlelight

goodnight

We do not always have light from the sun. Sometimes we use light from other sources. A light bulb is a light source. Fire is a light source, too.

Light is used in many ways. Plants need light to live and grow. Plants use light energy from the sun to make food.

Animals use light, too. The cat in the bottom picture is using light to find food.

Hurry! I'll save you!

People have many uses for light. We use light for working and playing. We use light for safety. How is light being used in each picture?

What would your life be like if you did not have many sources of light?

I know! I know!

Eric Brightspark

Einsteins' Checkup Time No. 1

1   Which things are light sources?

2   **True or false?**
   a) Plants don't need light to live and grow.
   b) We have all kinds of uses for light.
   c) A source of light is anything that gives off light energy.

3   **Think!**
Imagine you're a tiny animal like the golden mole, that always lives underground. There is no light for you to see by. If you can't see, how will you get around? How will you find your food?

Let's Stop For Another Chat

Do you know...
- how bright light and dim light affect what we see and how?
- which colours show up best in the dark?
- how light helps you to see?

Chapter 2

How do we see in bright and dim light?

bright light

Some sources give off a lot of light energy. They make a **bright** light. Other sources make only a **dim** light.

In bright light, we can see many colours. In dim light, it is hard to see dark colours.

dim light

Yellow and orange and white are easy to see in bright light and in dim light. We use these colours for safety.

Eric goes fluorescent.

If you go out walking or bicycling at night, it is especially good to wear light-coloured clothes. These will show up really well in a car's headlights.

Eric feels safe.

Some vehicles are painted in light colours too, for safety reasons. Why is this?

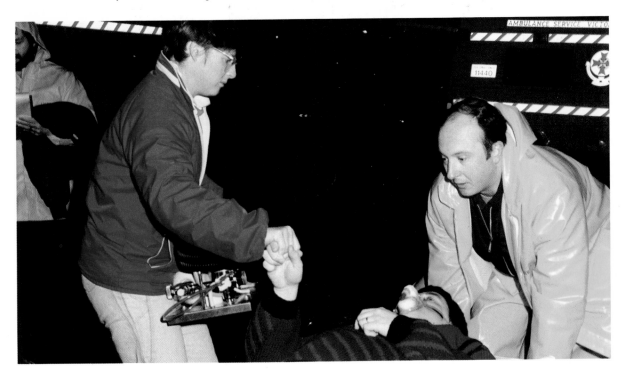

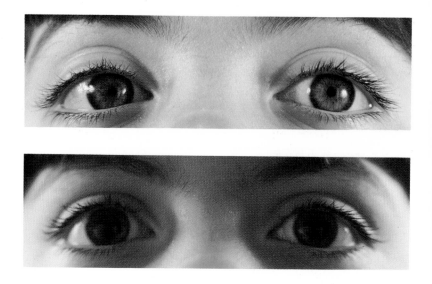

When light enters our eyes, we can see. Light goes into a small hole in the centre of the eye. This hole is called the **pupil**. The pupil looks like a small black circle.

The size of the pupil can change. In bright light, the pupil gets smaller. It lets in less light. When the light is dim, the pupil gets larger. It lets in more light.

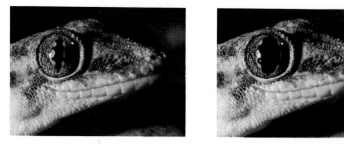

Many animals have pupils that can change size, too.

It is important to take care of your eyes. A doctor can check your eyes.

Never look straight into the sun or other bright light sources. You could hurt your eyes.

The people in the pictures are keeping their eyes safe.

Cool huh?

Very cool! Very funny!

Cool dude discovers shades.

Einsteins' Checkup Time No. 2

Hmmm! Interesting!

1 True or false?

a) It's better to wear black clothes for safety at night.

b) In bright light your pupils get bigger.

c) Yellow and orange show up better than other colours at night.

2 Which colour is safest in dim light?

3 Think! Why do cat-burglars, Ninja warriors and soldiers, blacken their faces when they go out on missions at night!

News Flash!

A Super Light

Have you ever seen a **laser** show? A laser is a special light. It is very different from other kinds of light. In a laser show, this light is used to make pictures.

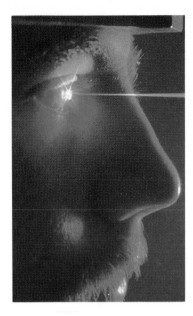

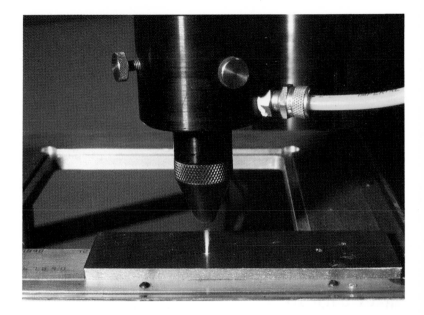

WOW!

Think of a light bulb. Its light spreads out in all directions. But a laser makes a thin beam of light. The light does not spread out. The beam is strong and straight and steady. It can travel a long, long way.

Lasers have many important uses. They can cut through steel. They can measure distances. They can help doctors. And new uses are being found all the time.

Think About It

If you had a laser, what could you measure with it?